The
Application
of
Mechanics
to Geometry

Popular Lectures in Mathematics

Survey of Recent East European Mathematical Literature

A project conducted by
Izaak Wirszup,
Department of Mathematics,
the University of Chicago,
under a grant from the
National Science Foundation

B. Yu. Kogan

The Application of Mechanics to Geometry

Translated and adapted from the Russian by David J. Sookne and Robert A. Hummel

The University of Chicago Press
Chicago and London

The University of Chicago Press, Chicago 60637
The University of Chicago Press, Ltd., London

International Standard Book Number: 0–226–45016–3
Library of Congress Catalog Card Number: 73–89789

Contents

1

The Composition
of Forces

1.1. Principal Assumptions

In this chapter we shall prove several theorems of geometry using the fundamental concepts and certain laws of statics. We will define the terms immediately.

1. *Force* is a vector and is characterized by magnitude, a direction, and a point of application. The line along which a force acts is called its *line of action*.

2. A body which cannot be deformed—that is, which always keeps its size and shape—is said to be *absolutely rigid*.

Actually, every body is capable of being deformed to some extent, but these deformations are frequently so small that they can be neglected. The concept of an absolutely rigid body is an idealization. One frequently omits the word *absolutely* and speaks simply of a *rigid body*.

3. A collection of forces acting on a body is called a *system of forces*. A system of forces is said to be *in equilibrium* or an *equilibrium system* if no motion is caused when the system is applied to an absolutely rigid body at rest.

4. Two systems of forces are said to be *equivalent* if they cause the same motion when applied to an absolutely rigid body.

From this definition it follows that, for all practical purposes, a system of forces acting on a rigid body can be replaced by any equivalent system without altering the discussion.

5. If a system of forces is equivalent to a single force **R**, we say that the force **R** is the *resultant* of this system.

Note that not every system of forces has a resultant. The simplest example of such a system of forces is called a *couple* of forces, as illustrated in figure 1.1.

In addition to the above concepts, we use the following rules (axioms) of statics:

1

RULE 1.1. *Two forces* $\mathbf{F}_1$ *and* $\mathbf{F}_2$ *acting at the same point have a resultant* $\mathbf{R}$ *which acts at the same point and is represented by the diagonal of the parallelogram having the forces* $\mathbf{F}_1$ *and* $\mathbf{F}_2$ *as adjacent sides* (fig. 1.2).

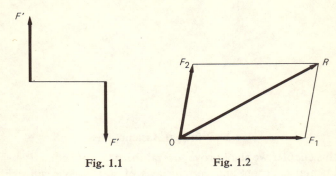

Fig. 1.1   Fig. 1.2

This construction is often called the *parallelogram law for forces*. The rule allows one to exchange the forces $\mathbf{F}_1$ and $\mathbf{F}_2$ for the force $\mathbf{R}$ and, conversely, to exchange a given force $\mathbf{R}$ for forces $\mathbf{F}_1$ and $\mathbf{F}_2$. In the first case one speaks of the *composition of forces*, and in the second, of the *resolution* of the force $\mathbf{R}$ into the components $\mathbf{F}_1$ and $\mathbf{F}_2$. (This resolution can be carried out in an infinite number of ways, since it is possible to construct infinitely many parallelograms with a given diagonal $\mathbf{R}$.)

RULE 1.2. *If we add any equilibrium system to a system of forces, or if we remove an equilibrium system from a system of forces, the resulting system will be equivalent to the original one.*

In particular, this implies that a collection of equilibrium systems is an equilibrium system.

RULE 1.3. *Two forces are in equilibrium if and only if they have the same magnitude, opposite directions, and a common line of action* (figs. 1.3 and 1.4).

Fig. 1.3   Fig. 1.4

RULE 1.4. *A force acting on a rigid body can be arbitrarily shifted along its line of action.*

In other words, if forces **F** and **F′** have the same magnitude and direction and a common line of action, they are equivalent (fig. 1.5). The converse is also true: If the forces **F** and **F′** are equivalent, they have the same magnitude and direction and a common line of action.[1]

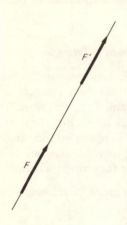

Fig. 1.5

Rule 1.4 implies that for forces acting on a rigid body, the point of application is unimportant; rather, the line of action determines the resultant force. The vector of a force acting on a rigid body is therefore called a *sliding vector*.

Rule 1.4 thus enables one to add forces whose points of application are different, providing their lines of action *intersect*. Suppose that we have to add forces F_1 and F_2 (fig. 1.6). Since the vectors of these forces are sliding, we can translate them to the point O and then use rule 1.1 to obtain the resultant **R** of the forces F_1 and F_2, by completing the parallelogram.

From rules 1.3 and 1.4 we deduce the following important proposition:

PROPOSITION 1.1. *If three nonparallel and coplanar forces acting on a rigid body are in equilibrium, then their lines of action intersect at a single point.*

For suppose that the forces P_1, P_2, and P_3 are in equilibrium with one another (fig. 1.7). Translating the forces P_1 and P_2 to the point O, we

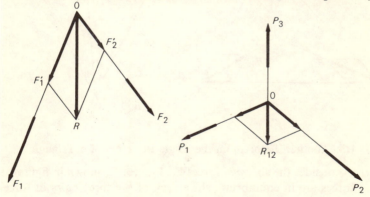

Fig. 1.6   Fig. 1.7

1. Rule 1.4 can be deduced from rule 1.3. We have not done this, however since both rules are equally intuitive.

obtain their resultant $\mathbf{R}_{12}$. The forces $\mathbf{P}_3$ and $\mathbf{R}_{12}$ are now in equilibrium. But this is possible only if they have a common line of action. Thus, the line of action of the force $\mathbf{P}_3$ passes through the point O—that is, the lines of action of all three forces meet one another at this point.

Using this proposition, we shall now prove some theorems of geometry.

1.2. A Theorem on the Angle Bisectors of a Triangle

Let us consider six equal forces $\mathbf{F}_1, \mathbf{F}_2, \ldots, \mathbf{F}_6$ acting along the sides of a triangle, as shown in figure 1.8. Since these forces cancel one another in pairs, they are clearly in equilibrium, and, therefore, the resultants $\mathbf{R}_{16}$, $\mathbf{R}_{23}$, and $\mathbf{R}_{45}$ are also in equilibrium. But the forces $\mathbf{R}_{16}$, $\mathbf{R}_{23}$, and $\mathbf{R}_{45}$ are directed along the bisectors of the interior angles A, B, and C. (The parallelograms are rhombi, and the diagonal is an angle bisector.) This leads, consequently, to the following theorem:

THEOREM 1.1. *The bisectors of the interior angles of a triangle intersect at a point.*

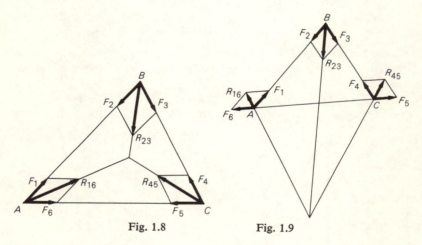

Fig. 1.8   Fig. 1.9

1.3. Another Theorem on the Angle Bisectors of a Triangle

Let us consider the six *equal* forces $\mathbf{F}_1, \mathbf{F}_2, \ldots, \mathbf{F}_6$ shown in figure 1.9. These forces are in equilibrium since each of the three pairs of forces, taken consecutively around the triangle, are in equilibrium. But the resultant of the forces $\mathbf{F}_1$ and $\mathbf{F}_6$ is directed along the bisector of the exterior angle A, and the resultant of $\mathbf{F}_4$ and $\mathbf{F}_5$ is directed along the bisector of interior angle C. The resultant of $\mathbf{F}_2$ and $\mathbf{F}_3$ is directed along the bisector of interior angle B. Therefore, the following theorem holds:

THEOREM 1.2. *The bisectors of two exterior angles and an interior angle of a triangle intersect at a point.*

1.4. A Theorem on the Altitudes of a Triangle

In figure 1.10 we have drawn a triangle ABC, with the forces $\mathbf{F}_1$, $\mathbf{F}_2$, ..., $\mathbf{F}_6$ acting along the sides. We have chosen these forces so that the following equalities hold:

$$
\begin{aligned}
F_1 &= F_2 = F \cos A \,, \\
F_3 &= F_4 = F \cos B \,, \\
F_5 &= F_6 = F \cos C \,,
\end{aligned}
\tag{1.1}
$$

where F is some convenient unit dimension for force. (Note that we are using the convention where an $\mathbf{F}$ denotes a vector, and F its corresponding magnitude.) Since the forces $\mathbf{F}_1$, $\mathbf{F}_2$, ..., $\mathbf{F}_6$ are in equilibrium, the lines of action of the resultants $\mathbf{R}_A$, $\mathbf{R}_B$, and $\mathbf{R}_C$ shown in the figure must intersect. We shall find the directions of these resultants.

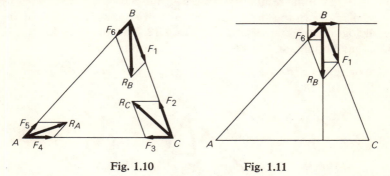

Fig. 1.10   Fig. 1.11

For example, let us add the forces $\mathbf{F}_1$ and $\mathbf{F}_6$, which act at the vertex B (fig. 1.11). To do this, we resolve each of these forces into two components, one parallel to the side AC, and the other perpendicular to it. The first of these components we shall call horizontal, and the second, vertical. From figure 1.11 it is clear that the horizontal components of the forces $\mathbf{F}_1$ and $\mathbf{F}_6$ are equal to $F_1 \cos C$ and $F_6 \cos A$. But from (1.1) it follows that

$$
\frac{F_1}{F_6} = \frac{\cos A}{\cos C} \,.
$$

Hence,

$$F_1 \cos C = F_6 \cos A \;.$$

Thus, the horizontal components of the forces $\mathbf{F}_1$ and $\mathbf{F}_6$ are the same. From this fact we conclude that they cancel one another and, therefore, the resultant of the forces $\mathbf{F}_1$ and $\mathbf{F}_6$ is perpendicular to side AC. Therefore, the force $\mathbf{R}_B$ is directed along the altitude perpendicular to AC.

Analogously, we may deduce that forces $\mathbf{R}_A$ and $\mathbf{R}_C$ lie along the two other altitudes of the triangle ABC. We thus arrive at the following:

THEOREM 1.3. *The altitudes of a triangle intersect at a single point.*

1.5. A Theorem on the Medians of a Triangle

Let us consider forces $\mathbf{F}_1$, $\mathbf{F}_2$, ..., $\mathbf{F}_6$, acting as shown in figure 1.12. Suppose that each of these forces has a magnitude equal to one-half the length of the corresponding side of the triangle. Then the resultant of the forces $\mathbf{F}_1$ and $\mathbf{F}_6$ will be represented by the median drawn to side BC; the resultant of forces $\mathbf{F}_2$ and $\mathbf{F}_3$ will be represented by the median drawn to side AC; and the resultant of forces $\mathbf{F}_4$ and $\mathbf{F}_5$ will be represented by the median drawn to side AB, since similar triangles are formed by the parallelograms of forces shown in figure 1.12. The forces $\mathbf{F}_1$, $\mathbf{F}_2$, ..., $\mathbf{F}_6$ are in equilibrium, and this leads to the following theorem:

THEOREM 1.4. *The medians of a triangle intersect at a single point.*

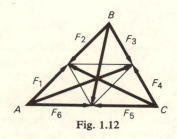
Fig. 1.12

1.6. A Generalization of the Theorem on the Bisectors of the Interior Angles of a Triangle

Suppose that we are given the triangle ABC. Let us draw a straight line a dividing angle A into parts α_1 and α_2, a straight line b dividing angle B into parts β_1 and β_2, and a straight line c dividing angle C into parts γ_1

and γ_2 (fig. 1.13). We apply to point A an arbitrary force $\mathbf{R}_1$ directed along the line a and resolve this force into components $\mathbf{P}_1$ and $\mathbf{Q}_1$ directed along the sides AC and AB. Similarly, we apply forces $\mathbf{R}_2$ and $\mathbf{R}_3$ directed along the lines b and c to the points B and C, and resolve these forces into components $\mathbf{P}_2$, $\mathbf{Q}_2$ and $\mathbf{P}_3$, $\mathbf{Q}_3$. We require, however, that the component $\mathbf{P}_2$ cancel the component $\mathbf{Q}_1$ and that the component $\mathbf{P}_3$ cancel the component $\mathbf{Q}_2$. In this way we obtain a system of forces $(\mathbf{R}_1, \mathbf{R}_2, \mathbf{R}_3)$ equivalent to the system $(\mathbf{P}_1, \mathbf{Q}_3)$.

Consider now the following ratios:

$$\frac{\sin \alpha_1}{\sin \alpha_2}, \quad \frac{\sin \beta_1}{\sin \beta_2}, \quad \frac{\sin \gamma_1}{\sin \gamma_2}.$$

From the parallelograms at the vertices A, B, and C, we deduce that

$$\frac{\sin \alpha_1}{\sin \alpha_2} = \frac{Q_1}{P_1},$$

$$\frac{\sin \beta_1}{\sin \beta_2} = \frac{Q_2}{P_2},$$

$$\frac{\sin \gamma_1}{\sin \gamma_2} = \frac{Q_3}{P_3},$$

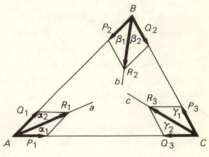

Fig. 1.13

and, therefore, that

$$\frac{\sin \alpha_1}{\sin \alpha_2} \frac{\sin \beta_1}{\sin \beta_2} \frac{\sin \gamma_1}{\sin \gamma_2} = \frac{Q_1}{P_1} \frac{Q_2}{P_2} \frac{Q_3}{P_3}.$$

But since $P_2 = Q_1$ and $P_3 = Q_2$,

$$\frac{\sin \alpha_1}{\sin \alpha_2} \frac{\sin \beta_1}{\sin \beta_2} \frac{\sin \gamma_1}{\sin \gamma_2} = \frac{Q_3}{P_1}. \tag{1.2}$$

Two cases are possible.

Case 1.

$$\frac{\sin \alpha_1}{\sin \alpha_2} \frac{\sin \beta_1}{\sin \beta_2} \frac{\sin \gamma_1}{\sin \gamma_2} = 1. \tag{1.3}$$

Then $P_1 = Q_3$. That is, the forces $\mathbf{P}_1$ and $\mathbf{Q}_3$ are in equilibrium; consequently the forces $\mathbf{R}_1$, $\mathbf{R}_2$, $\mathbf{R}_3$ which are equivalent to $\mathbf{P}_1$ and $\mathbf{Q}_3$, are in equilibrium. From this fact we then conclude that the lines a, b, and c intersect at a single point.

Case 2.

$$\frac{\sin \alpha_1}{\sin \alpha_2} \frac{\sin \beta_1}{\sin \beta_2} \frac{\sin \gamma_1}{\sin \gamma_2} \neq 1 \, .$$

Then, according to equation (1.2), $P_1 \neq Q_3$. We shall prove that in this case the lines a, b, and c cannot intersect at a single point. Suppose that they intersect at a single point O (fig. 1.14). Then, translating the forces $\mathbf{R}_1$, $\mathbf{R}_2$, and $\mathbf{R}_3$ to the point O, we find their resultant $\mathbf{R}$, which will also act at O. Furthermore, since the system $(\mathbf{R}_1, \mathbf{R}_2, \mathbf{R}_3)$ is equivalent to the system $(\mathbf{P}_1, \mathbf{Q}_3)$, the resultant $\mathbf{R}$ must be equivalent to the nonzero resultant of the forces $\mathbf{P}_1$ and $\mathbf{Q}_3$. This is impossible since the resultant of the forces $\mathbf{P}_1$, $\mathbf{Q}_3$ lies on the line AC, and the line of action of the force $\mathbf{R}$ cannot coincide with the line AC (since the point O does not lie on this line). From this contradiction, we conclude that the lines a, b, and c do not intersect at a single point.

Thus, the lines shown in figure 1.13 intersect at a single point only when equality (1.3) is valid. In other words, *the lines a, b, and c intersect at a single point if and only if condition* (1.3) *is satisfied.*

The theorem just proved may be regarded as a generalization of the theorem on the bisectors of the interior angles of a triangle. [In that theorem, not only does condition (1.3) hold, but also each of the individual factors ($\sin \alpha_1/\sin \alpha_2$, $\sin \beta_1/\sin \beta_2$, $\sin \gamma_1/\sin \gamma_2$) is equal to one.]

This theorem also implies the theorem on the altitudes of a triangle (fig. 1.15). If lines a, b, and c are drawn as altitudes, then

$$\frac{\sin \alpha_1}{\sin \alpha_2} = \frac{\cos C}{\cos B}, \qquad \frac{\sin \beta_1}{\sin \beta_2} = \frac{\cos A}{\cos C}, \quad \text{and} \quad \frac{\sin \gamma_1}{\sin \gamma_2} = \frac{\cos B}{\cos A} \, .$$

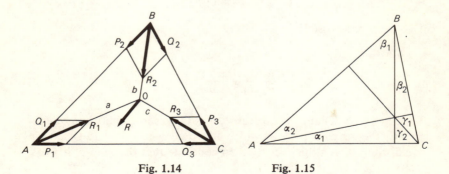

Fig. 1.14   Fig. 1.15

Taking the product, we find that

$$\frac{\sin \alpha_1}{\sin \alpha_2} \frac{\sin \beta_1}{\sin \beta_2} \frac{\sin \gamma_1}{\sin \gamma_2} = \frac{\cos C}{\cos B} \frac{\cos A}{\cos C} \frac{\cos B}{\cos A} = 1 \,.$$

Consequently, the altitudes of a triangle intersect at a single point.

1.7. Ceva's Theorem

Consider the triangle ABC (fig. 1.16). Suppose that forces $\mathbf{F}_1$ and $\mathbf{F}_2$ act along the sides AC and AB, and that their resultant acts along line AA_1. We draw the line DE parallel to side BC and resolve $\mathbf{F}_1$ into components $\mathbf{F}_1'$ and $\mathbf{F}_1''$, and $\mathbf{F}_2$ into components $\mathbf{F}_2'$ and $\mathbf{F}_2''$. From the similar triangles formed, it is evident that

$$\frac{F_1'}{F_1} = \frac{A_1C}{CA} \quad \text{and} \quad \frac{F_2'}{F_2} = \frac{BA_1}{AB} \,.$$

Therefore,

$$F_1' = F_1 \cdot \frac{A_1C}{CA} \quad \text{and} \quad F_2' = F_2 \cdot \frac{BA_1}{AB} \,.$$

But since the resultant of the forces $\mathbf{F}_1$ and $\mathbf{F}_2$ is directed along AA_1, $F_1' = F_2'$; consequently,

$$F_1 \cdot \frac{A_1C}{CA} = F_2 \cdot \frac{BA_1}{AB}$$

or

$$\frac{F_1}{F_2} = \frac{CA}{AB} \frac{BA_1}{A_1C} \,. \tag{1.4}$$

This relation will be needed later. (It is easy to remember because the right side of this equation can be obtained by circling the triangle CAB clockwise.)

Let us now determine A_1, B_1, and C_1 on triangle ABC (fig. 1.17). At the points A, B, and C we apply forces $\mathbf{R}_1$, $\mathbf{R}_2$, and $\mathbf{R}_3$ directed along lines AA_1, BB_1, and CC_1, and we resolve these forces into components directed along the sides of the triangle. The force $\mathbf{R}_1$ is chosen arbitrarily, but the forces $\mathbf{R}_2$ and $\mathbf{R}_3$ are chosen so that the equalities

$$P_2 = Q_1, \quad P_3 = Q_2 \tag{1.5}$$

are satisfied.

Applying relation (1.4) to each vertex, we have that

$$\frac{P_1}{Q_1} = \frac{CA}{AB} \frac{BA_1}{A_1C}, \qquad \frac{P_2}{Q_2} = \frac{AB}{BC} \frac{CB_1}{B_1A}, \qquad \frac{P_3}{Q_3} = \frac{BC}{CA} \frac{AC_1}{C_1B} \,.$$

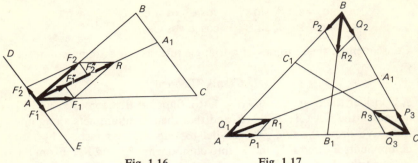

Fig. 1.16   Fig. 1.17

We next multiply these equalities to obtain

$$\frac{P_1}{Q_1}\frac{P_2}{Q_2}\frac{P_3}{Q_3} = \frac{BA_1}{A_1C}\frac{CB_1}{B_1A}\frac{AC_1}{C_1B},$$

or, canceling equals given in (1.5), and commuting,

$$\frac{AC_1}{C_1B}\frac{BA_1}{A_1C}\frac{CB_1}{B_1A} = \frac{P_1}{Q_3}. \tag{1.6}$$

We again consider two cases.

Case 1.

$$\frac{AC_1}{C_1B}\frac{BA_1}{A_1C}\frac{CB_1}{B_1A} = 1. \tag{1.7}$$

Then $P_1 = Q_3$, that is, these forces are in equilibrium. Consequently, the forces $\mathbf{R}_1$, $\mathbf{R}_2$, and $\mathbf{R}_3$ are in equilibrium; and, thus, the lines AA_1, BB_1, and CC_1 intersect at a single point.

Case 2.

$$\frac{AC_1}{C_1B}\frac{BA_1}{A_1C}\frac{CB_1}{B_1A} \neq 1.$$

Then according to (1.6) the forces $\mathbf{P}_1$ and $\mathbf{Q}_3$ are different. Repeating the argument of the preceding theorem, we deduce that the lines AA_1, BB_1, and CC_1 do not intersect at a point.

Thus the following theorem:

THEOREM 1.5. *For the lines AA_1, BB_1, and CC_1 to intersect at a single* *point, it is necessary and sufficient that equation* (1.7) *be valid. This result is known as Ceva's theorem.*[2]

The theorem on the medians of a triangle is a special case of Ceva's theorem, since, in the case for medians,

$$\frac{AC_1}{C_1B} = \frac{BA_1}{A_1C} = \frac{CB_1}{B_1A} = 1 \,.$$

In other words, Ceva's theorem can be considered a generalization of the theorem on the medians.

From Ceva's theorem it is also easy to obtain the previous theorem on the bisectors of the interior angles of a triangle. In this case

$$\frac{AC_1}{C_1B} = \frac{AC}{BC}\,, \qquad \frac{BA_1}{A_1C} = \frac{AB}{AC}\,, \qquad \frac{CB_1}{B_1A} = \frac{BC}{AB}\,,$$

and, consequently,

$$\frac{AC_1}{C_1B}\frac{BA_1}{A_1C}\frac{CB_1}{B_1A} = \frac{AC}{BC}\frac{AB}{AC}\frac{BC}{AB} = 1 \,.$$

Equality (1.7) is satisfied, and we may apply Ceva's theorem.

1.8. The Resultant and Its Point of Application

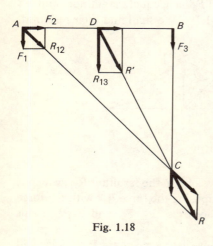

Fig. 1.18

We should make one more remark concerning the concept of the resultant. Suppose that the force **R** is the resultant of forces applied at different points of a rigid body. Since the vector **R** is a sliding vector, we can change its point of application by translating it along its line of action. But since the force **R** has no *actual* point of application (as it is not *directly* applied), any point on its line of action may be taken as its point of application. Thus, the resultant of forces applied at various points

2. By extending the lines forming the sides of the triangle, Ceva's theorem may be generalized to the case where the lines AA_1, BB_1, and CC_1 intersect *outside* the triangle ABC. The same is true, by the way, of the theorem of section 1.6.

of a rigid body has a definite *line* of action, but not a definite *point* of application.

To illustrate this statement, let us consider forces F_1, F_2, and F_3 as shown in figure 1.18. To find their resultant, we first add the forces F_1 and F_2 and then add their resultant R_{12} to the force F_3. In this way we finally end up with a resultant R applied at the point C. We now proceed in another way: First we add the forces F_1 and F_3, and then we add their resultant R_{13} to the force F_2. We then obtain a resultant R' acting at the point D. Thus, different methods of adding the forces F_1, F_2, and F_3 give resultants with different points of application. (One can assert, however, that the forces R and R' have the same line of action and that $R = R'$.)

From this argument we deduce the following rule:

RULE 1.5. *Suppose that by adding forces in various orders we obtain different points of application for their resultant. Then these points will be collinear, and the line formed will coincide with the line of action of the resultant.*

We shall now use this rule to prove two theorems.

1.9. A Third Theorem on the Angle Bisectors of a Triangle

Suppose that the forces F_1, F_2, and F_3 have equal magnitudes and act along the sides of triangle ABC (fig. 1.19). We shall find their resultant.

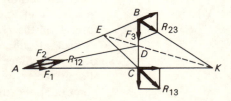

Fig. 1.19

Composing the forces F_1 and F_2, we obtain the resultant R_{12}, which is directed along the bisector AD. Composing the force R_{12} with the force F_3, we then find the resultant of the forces F_1, F_2, and F_3, and this resultant will act at the point D.

If we add the forces F_1 and F_3 first, we get the resultant R_{13}, which will lie on the extension of the bisector CE. Next we add the forces R_{13} and F_2, and again obtain the resultant of the forces F_1, F_2, and F_3. This time, however, the resultant acts at the point E.

Suppose we combine the forces $\mathbf{F}_2$ and $\mathbf{F}_3$ first. We then have the resultant $\mathbf{R}_{23}$, which will lie along the bisector BK of the exterior angle B. Composing the forces $\mathbf{R}_{23}$ and $\mathbf{F}_1$, we obtain a resultant which acts at the point K.

Thus, adding the forces $\mathbf{F}_1$, $\mathbf{F}_2$, and $\mathbf{F}_3$ in three different ways, we obtain resultants acting at the points D, E, and K. Consequently, the points D, E, and K are collinear. By defining the *base* of an angle bisector to be its point of intersection with the opposite side, we have the theorem:

THEOREM 1.6. *The bases of the bisectors of two interior angles and one exterior angle of a triangle form a straight line.*[3]

1.10. A Fourth Theorem on the Angle Bisectors of a Triangle

By carrying out a similar argument for three forces of equal magnitude, $\mathbf{F}_1$, $\mathbf{F}_2$, and $\mathbf{F}_3$, acting as shown in figure 1.20, we may then prove the theorem:

THEOREM 1.7. *The bases of the bisectors of the three exterior angles of a triangle are collinear* (fig. 1.21).

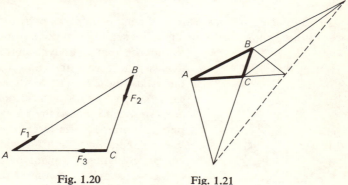

Fig. 1.20   Fig. 1.21

3. We assume that the bisector of the exterior angle intersects the opposite side —that is, is not parallel to this side. This remark is also applicable to the next theorem.

2

The Perpetual Motion Postulate

It is possible to prove certain geometric theorems using the postulate that perpetual motion is impossible. This chapter will demonstrate several theorems of this sort.

2.1. The Moment of Force

In addition to the postulate on the impossibility of perpetual motion, we will need the law of moments. We shall first state this law.

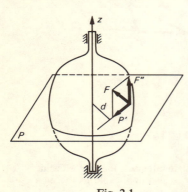

Fig. 2.1

Suppose that a body is under the influence of a force **F** and can revolve about the z-axis (fig. 2.1). The rotational motion caused by a force **F** is determined by its *moment* with respect to the z-axis. To compute this moment, we resolve the force **F** into components **F′** and **F″**, with the first component lying in a plane perpendicular to the z-axis, and the second parallel to the axis. The rotational motion caused by the component **F″** is clearly equal to zero, and the rotational action of the component **F′** is measured by the product of vector **F′** by the scalar d, where d is the distance between the z-axis and the line of action of the force **F′**. This product, denoted by **F′**d, is sometimes referred to as the *torque*, and in this text (for clarity) will be called the *moment of the force* **F** *with respect to the z-axis*.

Since the force **F′** is the projection of the force **F** onto the plane P, we can give the following definition of moment:

DEFINITION 2.1. *The moment of the force* **F** *with respect to the z-axis is the product* **F**$'d$, *where* **F**$'$ *is the projection of the force* **F** *onto the plane perpendicular to the z-axis, and d is the distance between the z-axis and the line of action of the projection* **F**$'$.

Thus,

$$M_z(\mathbf{F}) = \mathbf{F}'d \,,$$

where $M_z(\mathbf{F})$ is the moment of the force **F** with respect to the z-axis.

It follows from the definition that the moment of force is equal to zero in only two cases: when the line of action of the force **F** intersects the z-axis, or when it is parallel to the axis.

If, as frequently occurs, the force **F** has a line of action that lies in a plane perpendicular to the z-axis, then $F' = F$ and, therefore,

$$M_z(\mathbf{F}) = \mathbf{F}d \,.$$

In this case the distance d is called the *arm of the force* **F**.

We assign a definite sign to the moment of force. For this purpose we designate one of the directions of rotation as positive, and the other as negative. Then if the force tends to rotate the body in the positive direction, we consider its momentum positive, and in the opposite case, negative. Therefore, we can write

$$M_z(\mathbf{F}) = \pm \mathbf{F}'d \,,$$

where the sign is determined by the direction of rotation.

The following two rules will be needed:

RULE 2.1. *If* **R** *is the resultant of the system* $(\mathbf{F}_1, \mathbf{F}_2, \ldots, \mathbf{F}_n)$, *the moment of force* **R** *is equal to the vector sum of the individual moments of forces* $\mathbf{F}_1, \mathbf{F}_2, \ldots, \mathbf{F}_n$.[1]

This rule may be written in the form

$$M_z(\mathbf{R}) = M_z(\mathbf{F}_1) + M_z(\mathbf{F}_2) + \cdots + M_z(\mathbf{F}_n) \,, \tag{2.1}$$

where $M_z(\mathbf{R})$ denotes the moment of the force **R** with respect to the z-axis.

---

1. This proposition is known as *Varignon's theorem*. Varignon's name is also given to the theorem about segments joining midpoints of the sides of a quadrilateral.

RULE 2.2. (*The law of moments.*) *Suppose that a rigid body can rotate about a fixed axis. In order for forces acting on it not to cause rotation, it is necessary and sufficient that the vector sum of their moments equal zero.*

(In other words, the moment of the forces tending to rotate the body in the positive direction must have the same magnitude as the moment of the forces tending to rotate it in the negative direction.)

2.2. A Theorem on the Perpendicular Bisectors of the Sides of a Triangle

Consider a container having the form of a right triangular prism $A_1B_1C_1A_2B_2C_2$ (fig. 2.2). Imagine that it is filled with gas and that no external forces, not even gravity, act on it. (We can assume, for example, that it is located far from the earth and the heavenly bodies.) If we consider all the forces exerted on the container, we conclude that the net resultant must be constant, since no external force is applied. Since we have postulated that perpetual motion is impossible, this constant resultant must be zero. Thus, the container will remain in its initial state at rest. Consequently, the forces that the gas exerts on the walls must be in the equilibrium.

But since the pressures on the two parallel faces clearly balance one another, the forces exerted by the gas against the side walls of the container must be in equilibrium. We may represent these forces by $\mathbf{F}_{AB}$, $\mathbf{F}_{BC}$, and $\mathbf{F}_{AC}$, which lie in the plane defined by triangle ABC, and have points of application at the midpoints of their respective sides. Since these forces are in equilibrium, by our first proposition, their lines of action must intersect at a single point. Noting that the vectors $\mathbf{F}_{AB}$, $\mathbf{F}_{BC}$, and $\mathbf{F}_{AC}$ are perpendicular to the sides of the triangle ABC, we have the following:

THEOREM 2.1. *The perpendicular bisectors of the sides of a triangle intersect at a single point.*

2.3. The Pythagorean Theorem

Consider now a right triangular prism whose base is the right triangle ABC (fig. 2.3). We fill the container with gas and allow it to rotate about the vertical axis OO' (the ABC plane is considered horizontal). Since perpetual motion is impossible, the container will remain in its initial state at rest, and the forces caused by the gas on the side walls of the container must be in equilibrium. Each of these forces tends to rotate the container about the OO' axis: the forces $\mathbf{F}_1$ and $\mathbf{F}_2$ counterclockwise,

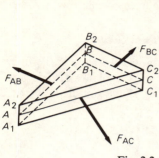

Fig. 2.2   Fig. 2.3

and the force F_3 clockwise. Therefore, the sum of the rotational moments of the forces F_1 and F_2 must equal the rotational moment of the force F_3. Since the arms of these forces are equal to $AB/2$, $BC/2$, and $AC/2$, respectively, we may use our formulas for rotational moments, and equate magnitudes, to obtain

$$F_1 \cdot \frac{AB}{2} + F_2 \cdot \frac{BC}{2} = F_3 \cdot \frac{AC}{2} . \tag{2.2}$$

But

$$F_1 = p(AB \cdot h) ,$$
$$F_2 = p(BC \cdot h) ,$$
$$F_3 = p(AC \cdot h) ,$$

where p is the pressure of the gas and h is the height of the container. Substituting, we find that equation (2.2) now takes the form

$$p(AB \cdot h) \frac{AB}{2} + p(BC \cdot h) \frac{BC}{2} = p(AC \cdot h) \frac{AC}{2} .$$

Multiplying by the constant $2/ph$, we have

$$AB^2 + BC^2 = AC^2 .$$

Thus, we have proved the following theorem:

THEOREM 2.2. *The sum of the squares of the legs of a right triangle is equal to the square of the hypotenuse.*

It is possible to generalize to the law of cosines by substituting an arbitrary triangle for the right triangle used in this proof. However, we shall now turn to a couple of geometric theorems dealing with circles.

2.4. A Theorem on Tangents and Secants

Suppose that a gas-filled vessel has a base whose shape is the figure *ABC* (figure 2.4 shows the view from above; the plane *ABC* is horizontal). The vessel has height *h* as measured along any of the vertical

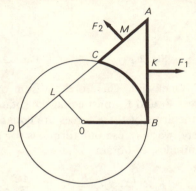

Fig. 2.4

sides joined to the lines *AB, AC,* or the arc *BC.* Thus the vessel is a boxlike container with cross-sectional shape *ABC.* Suppose, furthermore, that the vessel is tightly fastened to the rod *OB,* and that this rod is fastened to the vertical axis *O.* In this way, we allow the vessel to rotate about this axis. As in the preceding section, the vessel will remain at rest, and, therefore, the sum of the moments of all of the forces acting upon the container must be equal to zero. But only two of these forces create rotational moments: the forces F_1 and F_2 of the pressure of the gas on the walls *AB* and *AC.* (The forces of the gas on the curved wall *BC* do not contribute to the rotational moment, since each force has a line of action that passes through the axis *O.*) Using the fact that the moments of forces F_1 and F_2 have opposite signs and that the arms of these forces are equal to *BK* and *LM,* we know that

$$F_1 \cdot BK = F_2 \cdot LM .$$

But $BK = AB/2$ and

$$LM = \frac{LC + LA}{2} = \frac{LD + LA}{2} = \frac{AD}{2} .$$

Consequently,

$$F_1 \cdot \frac{AB}{2} = F_2 \cdot \frac{AD}{2} \,,$$

that is,

$$F_1 \cdot AB = F_2 \cdot AD \,. \tag{2.3}$$

Furthermore, we have that

$$F_1 = p(AB \cdot h) \quad \text{and} \quad F_2 = p(AC \cdot h) \,,$$

where p is the pressure of the gas and h the height of the vessel. Substituting these expressions in (2.3), we obtain

$$p(AB \cdot h)AB = p(AC \cdot h)AD \,,$$

hence,

$$AB^2 = AC \cdot AD \,. \tag{2.4}$$

Returning to the circle shown in figure 2.4, we see that the segment AB is tangent to the circle from point A; segment AD is a secant; and the segment AC is the external part of secant AD. Thus, equation (2.4) expresses the well-known geometric theorem:

THEOREM 2.3. *The square of the tangent to a circle from a point is equal to the product of a secant from that point times the external part of the secant.*

2.5. A Theorem on Two Intersecting Chords of a Circle

Let us replace the vessel just considered by the vessel ABD shown in figure 2.5. Repeating the arguments used in the beginning of the

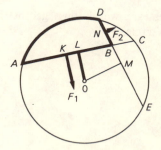

Fig. 2.5

preceding theorem, we obtain the relation

$$F_1 \cdot KL = F_2 \cdot MN \,.$$

Furthermore, since

$$KL = AL - AK = \frac{AC}{2} - \frac{AB}{2} = \frac{AC - AB}{2} = \frac{BC}{2},$$

$$MN = DM - DN = \frac{DE}{2} - \frac{DB}{2} = \frac{DE - DB}{2} = \frac{BE}{2},$$

therefore,

$$F_1 \cdot \frac{BC}{2} = F_2 \cdot \frac{BE}{2},$$

that is,

$$F_1 \cdot BC = F_2 \cdot BE.$$

As before, however,

$$F_1 = p(AB \cdot h), \quad F_2 = p(DB \cdot h),$$

and, consequently,

$$p(AB \cdot h) \cdot BC = p(DB \cdot h) \cdot BE,$$

hence,

$$AB \cdot BC = DB \cdot BE. \tag{2.5}$$

Equation (2.5) expresses a well-known geometric theorem on two intersecting chords of a circle.

3

The Center of Gravity, Potential Energy, and Work

In this chapter we shall compute the volumes and surfaces of certain bodies. Our discussion will make use of concepts involving potential energy, work, and properties of the center of gravity.

3.1. The Center of Gravity

The resultant of two parallel forces $\mathbf{F}_1$ and $\mathbf{F}_2$ (fig. 3.1) is represented by a force $\mathbf{R}$ acting in the same direction, with a line of action passing through point C. This point, which may be considered as the point of application of the resultant R, is defined by the relation

$$\frac{A_1C}{CA_2} = \frac{F_2}{F_1}. \tag{3.1}$$

Now let us consider a system of several parallel forces. For example, we shall find the resultant of four forces (fig. 3.2) by adding them one by one. Upon combining the forces $\mathbf{F}_1$ and $\mathbf{F}_2$, we obtain a force $\mathbf{R}'$

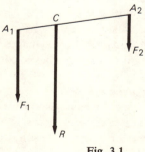

Fig. 3.1

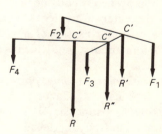

Fig. 3.2

acting at some point C', which may be determined by relation (3.1); adding the forces $\mathbf{R}'$ and $\mathbf{F}_3$, we obtain a resultant $\mathbf{R}''$ acting at a new point C''; and, finally, adding $\mathbf{R}''$ and $\mathbf{F}_4$, we have the resultant $\mathbf{R}$, acting at some point C whose position may be determined by repeated use of equation (3.1). The force $\mathbf{R}$ will have magnitude equal to $F_1 + F_2 + F_3 + F_4$.

In this manner, it is possible to add together any number of parallel forces to find their resultant and a point C at which the resultant acts. It can be proved that the position of C is independent of the order in which one adds the forces. This point is called the center of the given system of parallel forces.

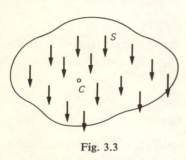

Fig. 3.3

Let us consider a rigid body located in the vicinity of the earth (fig. 3.3). If the dimensions of this body are small in comparison with the radius of the earth, the forces of gravity acting upon its particles are essentially parallel. We may therefore find a limit point C of a sequence of centers of arbitrarily large selections of parallel forces. This point is called the *center of gravity* of the given body. We may consider it to be the point of application of the force $\mathbf{P}$, which represents the weight of the body.

We should make a couple of observations that follow directly from this definition of the center of gravity. First, if we translate or rotate a body, the center point C is similarly translated; hence, relative to the body, the center of gravity remains fixed, or "firmly attached" to the body under rigid transformations.

Second, it follows from equation (3.1) that a proportional increase or decrease of all the forces in a system of parallel forces results in no change in the position of the center C. In other words, the center of gravity of a homogeneous body depends only on its size and shape. For this reason, the center of gravity is sometimes called the center of mass.

1. *The center of gravity of a line and a square.* In mechanics, one speaks of a material point, which is analogous to the geometric concept of a point.

A material point is a point possessing a definite mass.

The theoretical idea of a material point is a body having some measurable mass, but no dimensions. (Sometimes these concepts actually occur. For example, in studying the motion of the earth about the sun, we can treat the earth as a material point.)

Similarly, we can introduce the concept of a *material line*. By this we shall mean *a curve of finite length possessing a certain mass*. We shall represent this mass by its distribution along the length of the curve.

In the same manner, we can speak of a *material figure*. By this we shall mean a *plane figure possessing a definite mass distributed throughout its area*.

A material line may be visualized as a thin wire and a material figure as a thin plate. The thinner the wire or the plate, the closer it approximates the material line or material figure.

We may extend certain physical properties to our discussion of material lines and figures. By determining the weight of a material object per unit length, or per unit area, we may speak of the specific weight of the body. If the mass of a material curve is distributed uniformly along its length, the specific weight of this curve will be the same at all points. We shall call such a material curve *homogeneous*. In the same sense we can speak of a homogeneous material figure. A thin homogeneous wire *of constant diameter* serves as a prototype of a material curve. Analogously, a thin homogeneous plate *of constant thickness* is a prototype of a material figure.

Since a material curve and a material figure possess mass and consequently weight, one can speak of the center of gravity of a material line or a material figure. If the material curve or the material figure is homogeneous, the position of its center of gravity depends solely on its shape, and is independent of its specific weight. Therefore, the center of gravity of a homogeneous material curve can be called the *center of gravity of the curve*, and the center of gravity of a homogeneous material figure can be called the *center of gravity of the area* (or the *center of gravity of the figure*).

2. *The center of pressure*. It is not necessary to limit ourselves to forces of gravity when discussing the center of gravity of a thin plate.

Suppose that the pressure p acts on one side of a figure of area S (fig. 3.4). Since the forces associated with this pressure are parallel, their resultant is equal to

$$\mathbf{F} = pS,$$

and acts at some point C, which is the center of these forces. This point is, therefore, called the center of pressure. To determine its position, we must know the magnitude of the

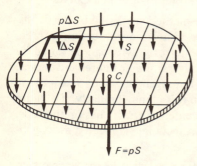

Fig. 3.4

force on any arbitrarily small area. But this force is equal to $p\Delta S$, where ΔS is the area of the cell on which the pressure acts (sometimes called a gaussian area). This is numerically equal to the weight of the cell, given that the material figure has a specific weight equal to p. Consequently, any partition of the figure yields forces of pressure having the same magnitudes as the forces of weight for a homogeneous material figure S. From this we may conclude that the point C coincides with the center of gravity of the figure S.

Thus, the center of a uniform pressure on a material figure is the same point as the center of gravity of the figure. This result will be needed in the proof of Guldin's first theorem.

3.2. Potential Energy

We shall assume the following propositions about potential energy in a gravitational force field:

PROPOSITION 3.1. *The potential energy of a material point is equal to Ph, where P is its weight and h its height.*

PROPOSITION 3.2. *The potential energy of a material system is equal to the sum of the potential energies of its points.*

PROPOSITION 3.3. *The potential energy of a rigid body is equal to Ph_C, where P is the weight of the body and h_C is the height of its center of gravity.*

The first two of these propositions are the definitions of the potential energy of a material point and a material system. It is important to note that the third proposition may be applied to material curves or material figures since these bodies have weight.

3.3. The Centers of Gravity of Certain Figures and Curves

To find the center of gravity using the method suggested by the definition, one must carry out the addition of a large number of parallel forces. In certain cases, however, the center of gravity can be found by an indirect method. We shall do this for several simple figures and curves.

1. *A rectangle.* It is known that if a homogeneous body has a plane of symmetry, its center of gravity lies on this plane. Similarly, if a figure or a curve has an axis of symmetry, its center of gravity lies on this axis. Consequently, the center of gravity of a rectangle is located at its *geometric center*.[1]

1. Consequently, the *center of pressure* of a rectangle is its geometric center. This fact was used several times in chapter 2 (for example, in the proof of Pythagoras' theorem).

2. *A circle.* By a similar argument, the center of gravity of a circle lies at its *center*.

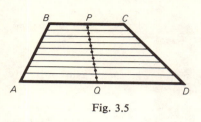

Fig. 3.5

3. *A triangular region.* Before generalizing to the case of a triangle, we will decompose a trapezoid into a large number of narrow strips of uniform width (fig. 3.5). Each strip has a center of gravity lying on the segment PQ which joins the midpoints of the bases AD and BC. Making the width of each strip infinitesimally small, we deduce that the center of gravity of the trapezoidal region lies on the line PQ.

Now suppose that the length of the upper base of the trapezoid converges to zero. Then the trapezoid approaches a triangle, and the line PQ becomes a median (fig. 3.6). Consequently, the center of gravity of a triangle lies on its median. But since this is true for each median of the triangle, its center of gravity C coincides with the point of intersection of its medians.

4. *A circular sector.* We now consider a circular sector ABO, viewed as a material figure lying in the vertical plane (fig. 3.7). Suppose that

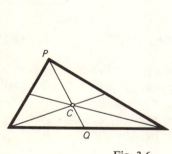

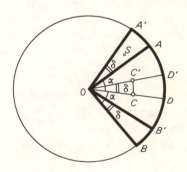

Fig. 3.6   Fig. 3.7

we have rotated the sector ABO about the center O by the angle δ, and that its new position is $A'B'O$. Let us calculate the change in its potential energy. Suppose that the point C is the center of gravity of the region ABO, and the point C' the center of gravity of the region $A'B'O$. The difference in potential energy between the positions $A'B'O$ and ABO is expressed by the formula

$$W_{A'B'O} - W_{ABO} = P_{ABO}H_{C'} \qquad (3.2)$$

where P_{ABO} is the weight of the sector and $H_{C'}$ is the height of the point C' above the horizontal line OD. (We have assumed that OC was originally horizontal.) But

$$H_{C'} = OC' \cdot \sin \delta = OC \cdot \sin \delta \quad \text{and} \quad P_{ABO} = \pi \cdot R^2 \frac{2\alpha}{2\pi} \cdot \gamma = R^2 \cdot \alpha \cdot \gamma,$$

where γ is the specific weight of the sector, and 2α measures the arc length of the sector, in radians. Therefore,

$$W_{A'B'O} - W_{ABO} = R^2 \alpha \gamma \cdot OC \cdot \sin \delta. \tag{3.3}$$

On the other hand,

$$W_{A'B'O} = W_{A'AO} + W_{AB'O} \quad \text{and} \quad W_{ABO} = W_{AB'O} + W_{BB'O}.$$

From these equations we get

$$W_{A'B'O} - W_{ABO} = W_{A'AO} - W_{BB'O}. \tag{3.4}$$

But since the regions $A'AO$ and $B'BO$ are congruent, the difference $W_{A'AO} - W_{B'BO}$ represents the change in potential energy of the sector $B'BO$ after being translated to the position $A'AO$. From the symmetry of the diagram, we have

$$W_{A'AO} - W_{BB'O} = 2H_S \cdot P_{A'AO} \tag{3.5}$$

where $P_{A'AO}$ is the weight of the sector $A'AO$, S is its center of gravity, and H_S is the height of the point S above the line OD. Furthermore, since

$$P_{A'AO} = \frac{R^2 \delta}{2} \gamma \quad (\delta \text{ is in radians})$$

and

$$H_S = OS \cdot \sin\left(\alpha + \frac{\delta}{2}\right),$$

equation (3.5) takes the form

$$W_{A'AO} - W_{B'BO} = R^2 \delta \gamma OS \cdot \sin\left(\alpha + \frac{\delta}{2}\right).$$

Combining this result with equation (3.4) and (3.3), we have

$$R^2\alpha\gamma\cdot OC\cdot\sin\delta = R^2\delta\gamma\cdot OS\cdot\sin\left(\alpha + \frac{\delta}{2}\right),$$

or,

$$OC = OS\cdot\frac{\sin\left(\alpha + \delta/2\right)}{\alpha}\cdot\frac{\delta}{\sin\delta}. \tag{3.6}$$

This equation allows us to compute OC. Since equation (3.6) is valid for arbitrary δ and, in particular, for δ as small as we please, we may write

$$OC = \lim_{\delta\to 0}\left[OS\cdot\frac{\sin\left(\alpha + \delta/2\right)}{\alpha}\cdot\frac{\delta}{\sin\delta}\right]. \tag{3.7}$$

But

$$\lim_{\delta\to 0}\frac{\sin\left(\alpha + \delta/2\right)}{\alpha} = \frac{\sin\alpha}{\alpha},$$

and from calculus we know that

$$\lim_{\delta\to 0}\frac{\delta}{\sin\delta} = 1.^2$$

Therefore, equation (3.7) becomes

$$OC = \left(\lim_{\delta\to 0}OS\right)\frac{\sin\alpha}{\alpha}. \tag{3.8}$$

Furthermore, as $\delta\to 0$ we can substitute the chord AA' for the arc $A'A$, and the triangle $A'AO$ for the sector $A'AO$. Therefore, as $\delta\to 0$, the point S approaches the point of intersection of the medians of this triangle. Since the medians of a triangle intersect at a point one-third of the distance from the base to the opposite vertex along the median, we may conclude that

$$\lim_{\delta\to 0}OS = \frac{2}{3}R.$$

2. This equation is contained in many textbooks on trigonometry (it is usually written in the form $\lim_{\delta\to 0}([\sin\delta]/\delta) = 1$). It expresses the intuitive fact that as an arc of a circle converges to zero, the ratio of its length to the length of the chord spanning it converges to unity.

Equation (3.8) now takes the form

$$OC = \frac{2}{3} R \cdot \frac{\sin \alpha}{\alpha} \, . \tag{3.9}$$

Formula (3.9) defines the position of the center of gravity of a circular sector.

5. *A half-disc.* Setting $\alpha = \pi/2$ in formula (3.9), we obtain

$$OC = \frac{4R}{3\pi} \, . \tag{3.10}$$

This equation defines the position of the center of gravity of a half-disc (fig. 3.8).

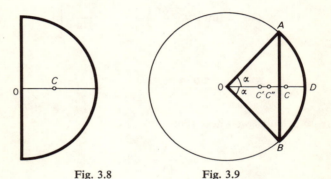

Fig. 3.8   Fig. 3.9

6. *A circular segment.* Suppose that a homogeneous material figure has the form of a circular sector and is located in a vertical plane (fig. 3.9). The sector $OADB$ is divided into the triangle OAB and the segment under investigation, ADB.

We allow the sector to rotate about the horizontal axis OD, and compute the moment of the forces of gravity acting upon it. Denoting the moments of the corresponding figures by M_{OADB}, M_{ADB}, and M_{OAB}, we may write

$$M_{OADB} = M_{ADB} + M_{OAB} \, . \tag{3.11}$$

But the moment of any system of forces and, in particular, of the forces of gravity, is equal to the moment of the resultant of the system (refer to rule 2.1; specifically, equation (2.1), on p. 15). We employ the formula $M_z(\mathbf{F}) = \mathbf{F}d$, to obtain

$$\begin{aligned}
M_{OADB} &= \gamma S_{OADB} \cdot OC'' \, , \\
M_{ADB} &= \gamma S_{ADB} \cdot OC \, , \\
M_{OAB} &= \gamma S_{OAB} \cdot OC' \, ,
\end{aligned} \tag{3.12}$$

where S is the area, γ is the specific weight, and C, C', and C'' are the centers of gravity of the segment, the triangle, and the sector, respectively. Substituting (3.12) into (3.11) and canceling γ, we have

$$S_{OADB} \cdot OC'' = S_{ADB} \cdot OC + S_{OAB} \cdot OC' . \tag{3.13}$$

Furthermore, we know that

$$S_{OAB} = R^2 \sin \alpha \cos \alpha, \qquad S_{OADB} = R^2 \alpha ,$$

$$OC' = \tfrac{2}{3} R \cos \alpha, \quad \text{and} \quad OC'' = \tfrac{2}{3} R \frac{\sin \alpha}{\alpha} .$$

In the last equation we have used formula (3.9). The others are derived from simple geometry. Relation (3.13) now takes the form

$$R^2 \alpha \left(\tfrac{2}{3} R \frac{\sin \alpha}{\alpha} \right) = S_{ADB} \cdot OC + R^2 \sin \alpha \cos \alpha \left(\tfrac{2}{3} R \cos \alpha \right) ,$$

or, simplifying,

$$S_{ADB} \cdot OC = \tfrac{2}{3} R^3 \sin \alpha (1 - \cos^2 \alpha) = \tfrac{2}{3} R^3 \sin^3 \alpha ,$$

$$OC = \frac{2 R^3 \sin^3 \alpha}{3 S_{ADB}} . \tag{3.14}$$

The equation obtained defines the position of the center of gravity of the segment. Since $R \sin \alpha = AB/2$, we may write our formula in the form

$$OC = \frac{(AB)^3}{12 S_{ADB}} ,$$

or, more briefly,

$$OC = \frac{l^3}{12 S} \tag{3.15}$$

where S is the area of the segment and l is its chord.

7. *An arc of a circle.* The center of gravity of a circular arc lies somewhere between the arc and the center of the circle defined by the arc. We may locate its position in the same way we found the center of gravity for a circular sector.

Let us rotate a homogeneous material arc AB lying in the vertical plane about the center point O by the angle δ (fig. 3.10). The new position of the arc is $A'B'$, and its center of gravity shifts from C to C'. The potential energy of the arc is increased by an amount

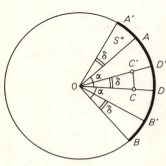

Fig. 3.10

$$W_{A'B'} - W_{AB} = P_{AB} H_{C'} , \tag{3.16}$$

where P_{AB} is the weight of the arc AB and $H_{C'}$ is the height of the point C' above the line OD. But

$$H_{C'} = OC' \cdot \sin \delta = OC \cdot \sin \delta \quad \text{and} \quad P_{AB} = 2R\alpha\gamma ,$$

where γ is the specific weight of the material arc. Therefore, equation (3.16) may be written in the form

$$W_{A'B'} - W_{AB} = 2R\alpha\gamma \cdot OC \cdot \sin \delta . \tag{3.17}$$

As before, we also have that

$$W_{A'B'} = W_{A'A} + W_{AB'}$$

and

$$W_{AB} = W_{AB'} + W_{B'B} ;$$

hence,

$$W_{A'B'} - W_{AB} = W_{A'A} - W_{B'B} . \tag{3.18}$$

But since the arcs $A'A$ and $B'B$ are congruent, the change in potential energy may be expressed as

$$W_{A'A} - W_{B'B} = 2P_{A'A}H_S$$

where $P_{A'A}$ is the weight of the arc $A'A$, S is its center of gravity, and H_S is the height of the point S above the line OD. However,

$$P_{A'A} = R\delta\gamma \quad \text{and} \quad H_S = OS \cdot \sin \left(\alpha + \frac{\delta}{2} \right) .$$

Therefore, from equation (3.18), our formula for potential energy becomes

$$W_{A'B'} - W_{AB} = 2R\delta\gamma \cdot OS \cdot \sin \left(\alpha + \frac{\delta}{2} \right) . \tag{3.19}$$

Equating (3.17) and (3.19), we have

$$2R\alpha\gamma \cdot OC \cdot \sin \delta = 2R\delta\gamma \cdot OS \cdot \sin \left(\alpha + \frac{\delta}{2} \right) ,$$

or

$$OC = OS \cdot \frac{\sin (\alpha + \delta/2)}{\alpha} \cdot \frac{\delta}{\sin \delta} .$$

Taking the limit as $\delta \to 0$, we obtain

$$OC = \left(\lim_{\delta \to 0} OS \right) \frac{\sin \alpha}{\alpha} . \tag{3.20}$$

But as $\delta \to 0$ the point S converges to A and, therefore,

$$\lim_{\delta \to 0} OS = R.$$

Therefore, substituting this value in (3.20), we find

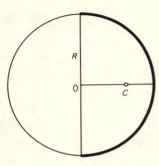

$$OC = R \cdot \frac{\sin \alpha}{\alpha}. \tag{3.21}$$

8. *A semicircle.* In the case when $\alpha = \pi/2$, formula (3.21) becomes

$$OC = \frac{2R}{\pi}. \tag{3.22}$$

Equation (3.22) defines the position of the center of gravity of a semicircle (fig. 3.11).

Fig. 3.11

We have now derived a number of formulas for the center of gravity of some simple figures. We will use these to compute surface area and volume of certain bodies.

3.4. The Volume of a Cylindrical Region

Suppose we generate a cylinder perpendicular to a closed curve lying in a plane. If another plane intersects the cylinder so as to bound a single region, we have a rather special cylindrical solid with at least one base perpendicular to the generator. We wish to find a formula for the volume of this type of solid.

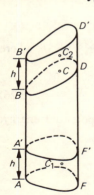

In figure 3.12, we have drawn a cylindrical solid of this type, with a vertical generator and a horizontal base. Suppose that we have lifted it by a certain height h, so that it now occupies the position $A'B'D'F'$. Let us calculate the change in potential energy.

Denoting the potential energy in the original position by W and in the final position by W', we have

$$W' - W = Ph,$$

where P is the weight of the solid and h the increase in the height of its center of gravity. Clearly,

Fig. 3.12

$$h = AA' = BB'.$$

Furthermore, supposing the solid to be homogeneous, we can write

$$P = V\gamma,$$

where V is the volume of the solid and γ its specific weight. Consequently,

$$W' - W = V\gamma h. \tag{3.23}$$

On the other hand,

$$W' = W_{A'BDF'} + W_{BB'D'D},$$
$$W = W_{A'BDF'} + W_{AA'F'F},$$

and, therefore,

$$W' - W = W_{BB'D'D} - W_{AA'F'F}, \tag{3.24}$$

that is, $W' - W$ is equal to the difference of the potential energy of the bodies $BB'D'D$ and $AA'F'F$. The volumes of these two regions are equal, and we may denote this volume by v. We will then have

$$W_{AA'F'F} = v\gamma H_{C_1},$$

$$W_{BB'D'D} = v\gamma H_{C_2}, \tag{3.25}$$

where C_1 and C_2 are the centers of gravity of the volumes $AA'F'F$ and $BB'D'D$, and H_{C_1} and H_{C_2} are the heights of the points C_1 and C_2 above the plane AF. Substituting the expressions (3.25) into equation (3.24), we get

$$W' - W = v\gamma(H_{C_2} - H_{C_1}).$$

Furthermore, since the body $AA'F'F$ is a right cylinder, $v = Sh$, where S is the area of the base AF. Therefore,

$$W' - W = Sh\gamma(H_{C_2} - H_{C_1}). \tag{3.26}$$

We now have two expressions for the change in potential energy. Setting them equal to one another, we obtain

$$V\gamma h = Sh\gamma(H_{C_2} - H_{C_1}),$$

or

$$V = S(H_{C_2} - H_{C_1}). \tag{3.27}$$

We are not finished, however, since equation (3.27) is valid for arbitrarily small h, and V does not depend on h. We proceed by taking a limit:

$$V = \lim_{h \to 0} [S(H_{C_2} - H_{C_1})] = S\left(\lim_{h \to 0} H_{C_2} - \lim_{h \to 0} H_{C_1}\right). \quad (3.28)$$

We shall compute the last two limits. First of all, it is clear that

$$\lim_{h \to 0} H_{C_1} = 0. \quad (3.29)$$

Furthermore, if h converges to zero, the points B' and D' converge to the points B and D, and the body $BB'D'D$ converges to a plate of constant thickness constructed on the base BD. Therefore, as $h \to 0$, the point C_2 converges to the center of gravity of a homogeneous material figure BD, or, in other words, to the center of gravity of the *figure BD*. Denoting this center of gravity by C, we get

$$\lim_{h \to 0} H_{C_2} = H_C, \quad (3.30)$$

where H_C is the height of the point C above the plane AF. Substituting (3.29) and (3.30) into (3.28), we now find

$$V = SH_C. \quad (3.31)$$

In other words, the volume of the solid is equal to the area of its base multiplied by the height of the center of gravity of the figure bounding the solid on top.

This relation will be useful for the computation of certain volumes.

We should make one further remark concerning the derivation of equation (3.31). In this derivation we have assumed that h can be chosen so small that the points of the figure $A'F'$ lie below the points of the figure BD. There are, however, cylindrical regions for which it is impossible to do this for any $h > 0$. This will be the case when the figure BD has a point in common with the base AF. An example of such a solid, labeled $ABDF$, is shown in figure 3.13. The proof given cannot be applied to this type of solid, and this case must be considered separately.

However, we may extend $ABDF$ to the solid $A'B'BDF'$ (fig. 3.14). We will then have

$$V_{ABDF} = V_{A'B'BDF'} - V_{A'B'BFF'} = S \cdot CC' - S \cdot C'C'' = S \cdot CC'', \quad (3.32)$$

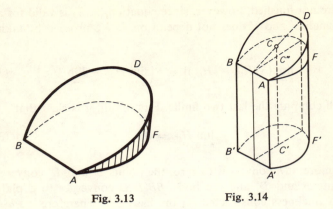

Fig. 3.13   Fig. 3.14

where $S = S_{ABF} = S_{A'B'F'}$, and C is the center of gravity of the figure ABD. But CC'' is the height of the point C above the plane ABF. Consequently, equation (3.32) becomes

$$V_{ABDF} = SH_C \, ,$$

where H_C denotes the height of the center of gravity C above the base of the solid $ABDF$. Thus, relation (3.31) is valid for cylindrical regions of the form shown in figure 3.13, as well as those previously considered.

Computing the volume of a solid by means of equation (3.31) poses a particular problem—we must locate the center of gravity of the upper surface. The following theorem may help make this task easier.

THEOREM 3.1. *Suppose that a cylinder is bounded by a surface perpendicular to the cylinder, with a center of gravity at C, and by a second surface with a center of gravity at C'. Then the line CC' will be parallel to the generator of the cylinder.*

Proof. Suppose that the line CC' is not parallel to the generator. Then the cylindrical region may be positioned so that the generator is horizontal and the line CC' inclined as shown for the solid $ABDF$ in

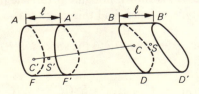

Fig. 3.15

figure 3.15. If H_C and $H_{C'}$ denote the heights of the points C and C' above some horizontal plane, then, by this construction,

$$H_C \neq H_{C'} . \tag{3.33}$$

Let us now displace the cylinder by a distance l in the direction of its generator. The new position will be $A'B'D'F'$, and since the shift is horizontal, the potential energy of the solid remains unchanged. Thus,

$$W_{A'B'D'F'} = W_{ABDF} . \tag{3.34}$$

But

$$W_{A'B'D'F'} = W_{A'BDF'} + W_{BB'D'D} ,$$

and

$$W_{ABDF} = W_{A'BDF'} + W_{AA'F'F} ,$$

and, therefore, equation (3.34) takes the form

$$W_{BB'D'D} = W_{AA'F'F} . \tag{3.35}$$

Using the notation as before, we have that

$$W_{BB'D'D} = P_{BB'D'D}H_S \quad \text{and} \quad W_{AA'F'F} = P_{AA'F'F}H_{S'}$$

where S and S' are the centers of gravity of the bodies $BB'D'D$ and $AA'F'F$, respectively. Since $P_{BB'D'D} = P_{AA'F'F}$, substituting these expressions into (3.35) yields

$$H_S = H_{S'} . \tag{3.36}$$

Now let l be very small. As l converges to zero, the body $BB'D'D$ will approach a homogeneous material figure BD, and the center of gravity C will converge to the center of gravity S of the figure BD. Consequently,

$$\lim_{l \to 0} H_S = H_C . \tag{3.37}$$

Similarly,

$$\lim_{l \to 0} H_{S'} = H_{C'} , \tag{3.38}$$

and from equations (3.36), (3.37), and (3.38) we conclude that

$$H_C = H_{C'} .$$

But since this result contradicts relation (3.33), our original assumption that the line CC' is not parallel to the generator of the cylindroid must be false.

This proves that the center of gravity of the upper base of a cylindrical solid lies directly above the center of gravity of the bottom base.[3]

We shall now compute the volume of two special cylindrical solids.

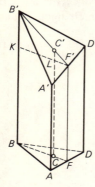

Fig. 3.16

1. We shall compute the volume of a prism, which may be viewed as a cylindrical region with a triangular base. Let C and C' denote the centers of gravity of its bases. Then, by equation (3.31), we will have

$$V = S \cdot CC', \qquad (3.39)$$

where S is the area of triangle ABD. Furthermore, since the center of gravity of a triangle is located at the point of intersection of its medians,

$$C'F' = \tfrac{1}{3}B'F',$$

and, consequently,

$$LC' = \tfrac{1}{3}KB'.$$

Therefore,

$$CC' = FF' + \tfrac{1}{3}(BB' - FF') = \frac{2FF' + BB'}{3}.$$

But since

$$FF' = \tfrac{1}{2}(AA' + DD'),$$

we have that

$$CC' = \frac{AA' + BB' + DD'}{3}.$$

Substituting this expression into (3.39), our formula becomes

$$V = S\frac{AA' + BB' + DD'}{3}. \qquad (3.40)$$

3. This proof is not applicable to the solid shown in figure 3.13. However, the same trick used in figure 3.14 allows us to claim the result for the general case.

In general, for a triangular prism with at least one of the two end faces perpendicular to the length of the body, if S is the area of the cross section, and H_1, H_2, and H_3 are the measures of the heights along the length of the body, then the desired formula for volume may be written as

$$V = S \frac{H_1 + H_2 + H_3}{3}. \tag{3.41}$$

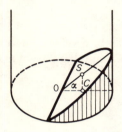

Fig. 3.17

2. We next consider a region obtained from a circular cylinder by a plane section passing through the diameter of the lower base (fig. 3.17). Its volume is equal to

$$V = \frac{\pi R^2}{2} \cdot CS = \frac{\pi R^2}{2} \cdot OC \cdot \tan \alpha$$

where C is the center of gravity of the half-disc which forms the base of the solid. But, as we found earlier,

$$OC = \frac{4R}{3\pi}$$

(see formula (3.10) and fig. 3.8). Consequently,

$$V = \frac{\pi R^2}{2} \frac{4R}{3\pi} \tan \alpha = \tfrac{2}{3} R^3 \tan \alpha. \tag{3.42}$$

It is interesting to note that this formula for volume does not use the value of pi.

3.5. The Volume of a Pyramid

Let us consider a triangular pyramid, one of whose sides is perpendicular to the plane of the base (fig. 3.18). Since this is merely a special case of the prism shown in figure 3.16, we may find its volume by using formula (3.41). Setting $H_1 = H_3 = 0$ and $H_2 = H$ in this formula, we have

$$V = \tfrac{1}{3} SH.$$

Now suppose that we have a pyramid with a more general base, that is, any arbitrary polygon lying in a plane (fig. 3.19). We may then divide

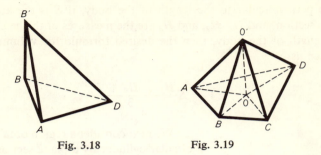

Fig. 3.18   Fig. 3.19

it into several triangular pyramids of the form shown in figure 3.18. The pyramid shown in figure 3.19, for example, can be divided into four such pyramids having the edge OO' in common. Computing the volume of each of these pyramids, we have

$$V_1 = \tfrac{1}{3}S_1H_1 ,$$
$$V_2 = \tfrac{1}{3}S_2H_2 ,$$
$$V_3 = \tfrac{1}{3}S_3H_3 ,$$
$$V_4 = \tfrac{1}{3}S_4H_4 .$$

Therefore,

$$V = V_1 + V_2 + V_3 + V_4 = \tfrac{1}{3}(S_1 + S_2 + S_3 + S_4)H ,$$

or,

$$V = \tfrac{1}{3}SH , \tag{3.43}$$

where S is the area of the base $ABCD$. Formula (3.43) is the familiar expression for the volume of a pyramid.[4]

3.6. The Volume of a Body of Revolution (Guldin's First Theorem)

Let us consider a body obtained by the rotation of a plane figure Q about an axis lying in its plane (half of such a body is shown in figure 3.20). We assume the axis OO' to be vertical. Let us construct a new structure by adjoining to the body shown in figure 3.20 a cylindrical pipe, and assume the entire pipe structure to be hollow and connected

4. We divided the pyramid shown in figure 3.19 into four pyramids of the type shown in figure 3.18. However, if a pyramid is very "oblique," the point O can lie outside of the base $ABCD$ and such a decomposition will not be possible. We will then need to consider the "algebraic" sum of pyramids rather than the "arithmetic sum" (that is, take the volumes of several of the pyramids with a negative sign).

by a narrow neck (figure 3.21 shows the top view; the axis OO' is represented by the point O). We insert a piston $ABDE$ in the cylindrical part of the pipe and the piston $KLMN$ in the circular part. We fill the cavity between the pistons with an incompressible liquid. Suppose that the force **F** acts on the piston $ABDE$, forcing it to assume the position $A'B'D'E'$. We compute the work performed by this force.

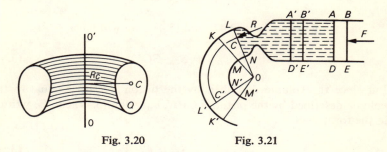

Fig. 3.20   Fig. 3.21

Since the path traveled by the point of application of the force **F** is equal to BB', the work is equal to

$$A = F \cdot BB' = F \cdot AA' . \tag{3.44}$$

The force **F** is equal to the pressure of the liquid on the wall AD. Therefore,

$$F = pS_{AD} , \tag{3.45}$$

where p is the pressure of the liquid and S_{AD} is the area of the piston face AD. Substituting this expression in (3.44), we have

$$A = pS_{AD} \cdot AA' , \tag{3.46}$$

and since $S_{AD} \cdot AA'$ represents the volume of the section $A'AD'D$,

$$A = pV_{A'AD'D} . \tag{3.47}$$

This is the expression for the work performed by the force **F**.

Let us now compute this work in another way. We shall consider it to be the work performed by the force **R** pressing against the piston $KLMN$. We then get

$$A = R \cdot CC' ,$$

where C is the point of application of the force $\mathbf{R}$ and CC' is the length of the arc described by this point during the motion of the piston. But

$$R = pS_{LN},$$

where S_{LN} is the area of the piston-face LN. Consequently,

$$A = pS_{LN}\cdot CC'. \tag{3.48}$$

Comparing expressions (3.47) and (3.48), we will have

$$V_{A'AD'D} = S_{LN}\cdot CC'. \tag{3.49}$$

But since the volume described by the piston $ABDE$ is equal to the volume described by the piston $KLMN$, equation (3.49) can be written in the form

$$V_{L'LNN'} = S_{LN}\cdot CC'. \tag{3.50}$$

Note that S_{LN} is simply the area of the figure Q in figure 3.20, and $V_{L'LNN'}$ is the volume of the region obtained by the rotation of this figure. Since formula (3.50) is valid for any arc length CC', in general,

$$V = S\cdot CC' \tag{3.51}$$

where V is the volume of the region generated, and S is the area of the figure Q.

The value of CC' is determined by the length of the arc described by the point C as the figure is revolved. But the point C has a simple geometric interpretation. Since it is the center of pressure of the wall LN and the pressure is uniform at all points of this wall, the point C coincides with the center of gravity of the figure Q. Consequently, the arc CC' is the arc described by the center of gravity of this figure.

Let us apply equation (3.51) to the body drawn in figure 3.20 (half of a body of rotation). In this case the arc CC' is equal to πR_C, where R_C is the distance of the point C from the axis of rotation. Consequently, the volume of the body under consideration is equal to $S\pi\cdot R_C$. But since this volume is half that of the complete body of rotation, we can write

$$\frac{V}{2} = S\pi R_C,$$

hence,

$$V = 2\pi R_C S . \qquad (3.52)$$

We now have the following theorem:

THEOREM 3.2. *The volume of a body of revolution is equal to the area of the figure from which it is obtained multiplied by the length of the circumference of the circle described by the center of gravity of this figure.*

This theorem is known as Guldin's first theorem.[5]

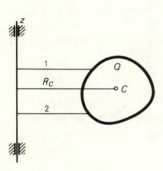

Fig. 3.22

Another proof of Guldin's first theorem. Suppose that the plate Q is free to rotate about the horizontal z-axis (figure 3.22 shows the view from above). We shall assume that Q is a material figure of weight P. The plane of the plate is horizontal and passes through the z-axis. Rods 1 and 2 are assumed weightless.

The weight of the plate creates a rotational moment with respect to the z-axis. This moment is equal to

$$M_z(P) = P R_C , \qquad (3.53)$$

where R_C is the distance from the center of gravity of the figure Q to the z-axis. But

$$P = \gamma S , \qquad (3.54)$$

where S is the area of the figure Q and γ is its specific weight. (We assume the figure is homogeneous.) Consequently,

$$M_z(P) = \gamma S R_C . \qquad (3.55)$$

The product $S R_C$ is called the static moment of the area S with respect to the z-axis. We shall denote it by $M_z(S)$, that is,

$$M_z(S) = S R_C . \qquad (3.56)$$

5. Paul Guldin (1577–1643), Swiss mathematician known for introducing the centrobaric method.

Before proceeding with the proof, we should investigate this new term carefully. Comparing equations (3.53) and (3.56), we see that formula (3.56) is obtained from formula (3.53) by the substitution of S for P. Loosely speaking, therefore, the static moment of area is the moment "created by the area" of the figure under consideration. Moreover, from equations (3.55) and (3.56), it is clear that the static moment of area can be viewed as the moment created by the weight of a figure for which $\gamma = 1$.

From equations (3.55) and (3.56) it follows that

$$M_z(P) = \gamma M_z(S) . \tag{3.57}$$

This relation serves as a conversion formula for the moment of weight and the moment of area.

Let us divide the plate into several parts. We will then be able to write

$$M_z(P) = M_z(P_1) + M_z(P_2) + \cdots + M_z(P_n) , \tag{3.58}$$

where $P_1, P_2, \ldots, P_n$ are the weights of these parts. (Note that we have used the rule that the moment of a resultant is equal to the sum of the moments of the forces of the system.) Using the conversion formula (3.57) for each moment, we obtain

$$\gamma M_z(S) = \gamma M_z(S_1) + \gamma M_z(S_2) + \cdots + \gamma M_z(S_n) .$$

Finally, dividing this equation by γ (or setting $\gamma = 1$), we arrive at the following rule:

. RULE 3.1. *If the area S consists of the areas $S_1, S_2, \ldots, S_n$, then*

$$M_z(S) = M_z(S_1) + M_z(S_2) + \cdots + M_z(S_n) . \tag{3.59}$$

We shall now use this equation for the proof of Guldin's theorem.

Suppose in figure 3.22 that the plate Q is a rectangle with one of its sides parallel to the z-axis (fig. 3.23). The volume of the body of rotation which is obtained is expressed by

$$V = \pi R_2^2 h - \pi R_1^2 h ,$$

or

$$V = 2\pi(R_2 - R_1)h \frac{R_2 + R_1}{2} . \tag{3.60}$$

But

$$(R_1 - R_2)h = S,$$

and

$$\frac{R_1 + R_2}{2} = R_C,$$

where S is the area of the rectangle and C its center of gravity. Consequently,

$$V = 2\pi S R_C,$$

that is,

$$V = 2\pi M_z(S), \tag{3.61}$$

where $M_z(S)$ is the moment of area of the rectangle with respect to the axis of rotation.

Let us now substitute an arbitrary figure Q for this rectangle (fig. 3.24). We divide this figure into a large number of narrow strips and

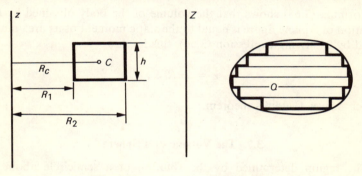

Fig. 3.23   Fig. 3.24

approximate each of these strips by the rectangle inscribed in each strip. If n denotes the number of strips, and we allow this number without bound, the approximations become successively better. We then have

$$V = \lim_{n \to \infty} (V_1 + V_2 + \cdots + V_n),$$

where V is the volume of the body obtained by the rotation of the figure Q and $V_1, V_2, \ldots, V_n$ are the volumes of the bodies obtained by the rotation of each of the rectangles. But according to (3.61),

$$V_1 = 2\pi M_z(S_1)$$
$$\vdots$$
$$V_n = 2\pi M_z(S_n).$$

Consequently,

$$V = 2\pi \cdot \lim_{n \to \infty} [M_z(S_1) + M_z(S_2) + \cdots + M_z(S_n)] \qquad (3.62)$$

where $S_1, S_2, \ldots, S_n$ are the areas of the rectangles. But according to (3.59) the sum inside the square brackets is equal to the static moment of the area of a cross section (the area bounded by the boldface lines in figure 3.24). But since this figure converges to the figure Q as $n \to \infty$,

$$\lim [M_z(S_1) + M_z(S_2) + \cdots + M_z(S_n)] = M_z(S),$$

where $M_z(S)$ is the static moment of the area bounded by the figure Q. Therefore, equation (3.62) takes the form

$$V = 2\pi M_z(S). \qquad (3.63)$$

Formula (3.63) shows that the volume of the body obtained by the rotation of a plane figure is equal to the static moment of its area multiplied by 2π. Using expression (3.56), this implies

$$V = 2\pi R_c S,$$

which proves Guldin's theorem.

3.7. The Volume of a Sphere

The region determined by the rotation of a semicircle about its diameter forms a sphere. Substituting into the formula given above, the volume of the sphere may be expressed as

$$V = 2\pi \cdot OC \frac{\pi R^2}{2},$$

where C is the center of gravity of the semicircular plate (see figure 3.8). But according to formula (3.10)

$$OC = \frac{4R}{3\pi},$$

and, consequently,

$$V = 2\pi \cdot \frac{4R}{3\pi} \cdot \frac{\pi R^2}{2},$$

that is,

$$V = \frac{4}{3}\pi R^3 .$$

3.8. The Volumes of Certain Other Bodies of Rotation

The strength of Guldin's theorem lies in its applicability to a large number of different solids. We will give several further examples.

1. *A circular cylinder.* Referring to figure 3.25, it is clear that the volume of a circular cylinder may be given by

$$V = 2\pi R_C S = 2\pi \frac{R}{2} RH = \pi R^2 H ,$$

where R is the radius of the cylinder and H is its altitude.

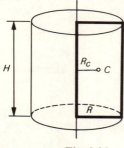   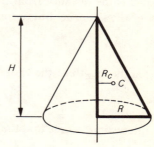

     Fig. 3.25   Fig. 3.26

2. *A cone.* A circular cone can be viewed as a body obtained by the rotation of a right triangle about one of its legs (fig. 3.26). The center of gravity of a triangle is located at the point of intersection of its medians. Consequently, R_C is equal to $R/3$, and we obtain the familiar expression for the volume of a cone:

$$V = 2\pi R_C S = 2\pi \frac{R}{3} \frac{RH}{2} = \frac{\pi R^2 H}{3} .$$

3. *A torus.* A torus is a body obtained by the rotation of a circle about an axis lying in the same plane as the circle (fig. 3.27). In accordance with Guldin's theorem, the volume of a torus is equal to

$$V = 2\pi R_C S = 2\pi R \pi r^2 = 2\pi^2 R r^2 .$$

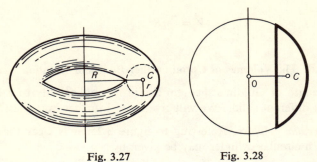

Fig. 3.27   Fig. 3.28

4. Suppose that a circular segment rotates about the diameter parallel to its chord (fig. 3.28). The volume of the ring-shaped body obtained is equal to

$$V = 2\pi \cdot OC \cdot S \,,$$

where OC is the distance from the center of the circle to the center of gravity of the segment. But according to (3.15),

$$OC = \frac{l^3}{12S} \,,$$

where l is the length of the chord of the segment. Therefore,

$$V = \frac{\pi l^3}{6} \,.$$

It is interesting to note that the volume which we have found depends only on l.

We have, by no means, exhausted the applications of Guldin's theorem. For example, figure 3.29 shows a solid obtained by the rotation of a circular segment about its chord. Since we know the location of the center of gravity of a segment, the reader might want to determine the volume of this solid.

5. Guldin's theorem is often useful to substantially reduce the number of necessary calculations. Suppose, for example, that the square $ABDE$ can rotate about the axis OO' (fig. 3.30). The volume of the corresponding solid of rotation can be found by computing the difference between the volumes of the two inclined edges which form slices from a cone. This, however, is comparatively complicated, since by means of Guldin's theorem, we can immediately obtain

$$V = 2\pi R_C S = 2\pi \left(a + \frac{a}{2} \right) \frac{a^2}{2} = \frac{3}{2} \pi a^3 \,,$$

where a is the diagonal of the square.

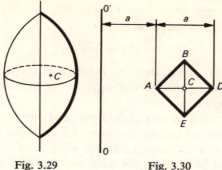

Fig. 3.29   Fig. 3.30

A similar example is illustrated by the following problem. A triangle rotates once about the axis z_1 and once about the axis z_2. Given that the axis z_2 is parallel to z_1 (fig. 3.31), how are the volumes of the corresponding bodies of rotation related?

Guldin's theorem allows one to solve this problem without carrying out any computations. Since the medians of the triangle intersect at a point which is twice as far from the axis z_2 as it is from z_1, $V_1 : V_2 = 1 : 2$.

6. Suppose that a homogeneous material figure Q lies in a horizontal plane (fig. 3.32). If we allow it to rotate about the horizontal axis OO' passing through its center of gravity, it will remain in equilibrium. Consequently,

$$\gamma S_1 R_1 = \gamma S_2 R_2 ,$$

where γ is the specific weight of the figure, S_1 and S_2 are the areas of the

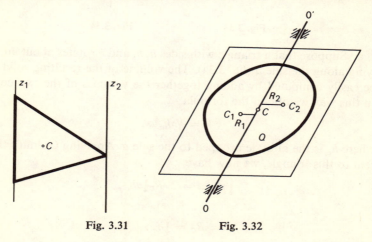

Fig. 3.31   Fig. 3.32

two parts divided by the line OO', and R_1 and R_2 are the distances of the centers of these parts from the line OO'. Multiplying the above equation by $2\pi/\gamma$, we will have

$$2\pi R_1 S_1 = 2\pi R_2 S_2 . \tag{3.64}$$

Equation (3.64) shows that the volumes of the two distinct solids obtained by rotating the left and right halves of this figure about the axis OO' are the same.

This result is valid for any plane figure and any straight line passing through its center of gravity. Suppose, for example, that the triangle ABC rotates about the median BD (fig. 3.33). Then the volumes of the bodies described by the triangles ABD and BDC are equal.

7. In all the examples up to this point, Guldin's theorem has been used strictly for the calculation of volumes. It is possible, however, to use it in another way. Knowing the volume of a body of revolution, we may find the center of gravity of the figure from which the body is obtained. Let us consider two examples.

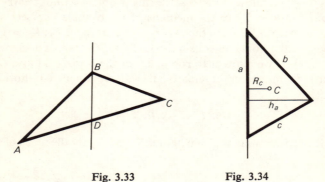

Fig. 3.33   Fig. 3.34

a. Suppose that a triangle with sides a, b, and c rotates about an axis lying along the side a (fig. 3.34). The volume of the resulting solid can be easily computed by adding together the volumes of the two cones. In this way we obtain the formula

$$V = \tfrac{1}{3}\pi h_a^2 a ,$$

where h_a is the altitude dropped to the side a. Applying Guldin's theorem to this triangle, we now have

$$\tfrac{1}{3}\pi h_a^2 a = 2\pi R_C \tfrac{1}{2} a h_a ,$$

or,

$$R_C = \tfrac{1}{3} h_a .$$

Thus, the center of gravity of this triangle lies at a distance $\frac{1}{3}h_a$ from side a. Arguing in the same way, we may deduce that it lies at distances $\frac{1}{3}h_b$ and $\frac{1}{3}h_c$ from sides b and c, respectively. But only one point of the triangle possesses this property—the point of intersection of its medians.

b. As a second example, we shall find the center of gravity of a half-disc. Applying Guldin's theorem to the sphere, we obtain

$$\frac{4}{3}\pi R^3 = 2\pi \cdot OC \cdot \frac{\pi R^2}{2},$$

where O is the center of the disc and C the center of gravity of the half-disc. Consequently,

$$OC = \frac{4R}{3\pi}. \tag{3.65}$$

Formula (3.65) defines the position of the center of gravity of a half-disc.

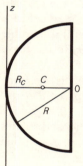

Fig. 3.35

Of course, if we derive formula (3.65) in this way, we may be accused of circular reasoning if we use the formula to compute the volume of a sphere. We can, however, use it to compute the volumes of certain other bodies, for example, the body shown in figure 3.17 (see equation (3.42) on p. 37). Thus, Guldin's theorem allows us to compute the volume of this body starting from the formula for the volume of a sphere. We can find other examples of this kind. Suppose, for example, that a half-disc rotates about the z-axis (fig. 3.35). The volume of the resulting body of rotation can be found by computing the difference $R_C = R - OC$, applying formula (3.65). Consequently, by assuming the formula for the volume, we can compute the volume of this body of rotation. (Note that this body is not the "sum" or "difference" of solids whose volumes are known. For this reason, a direct computation of its volume turns out to be difficult.)

3.9. The Surface of a Body of Rotation (Guldin's Second Theorem)

Let us first introduce two new concepts.

1. *The tangent.* Let us take points M and M' on the curve AB and draw the secant MM' (fig. 3.36). We now fix the point M and bring the

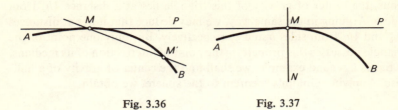

Fig. 3.36   Fig. 3.37

point M' arbitrarily close to M. The secant MM' will then approach, as a limit, the position MP. The line MP is called the *tangent to the curve AB at the point M*.

2. *The normal.* Let us take a point M on a plane curve (fig. 3.37). We draw the tangent MP and the line MN perpendicular to MP through the point M. The line MN is called the *normal* to the given curve at the point M. Loosely speaking, we may refer to it as the perpendicular to the curve AB at the point M.

Let us now consider the arc AB of a plane curve (fig. 3.38). We take an arbitary point C on the curve, draw the normal through this point, and mark off a small segment CC' of given length d along the normal. Drawing such segments at each point of the arc AB, we may define a curve $A'B'$ as the locus of the endpoints of these segments. It is possible to prove that each of the segments CC' is normal not only to the curve AB, but also to the curve $A'B'$. Therefore, the distance CC' can be viewed as the width of the strip $AA'B'B$. Since this width is the same at all points along the curve, we will say that the strip $AA'B'B$ has *constant width*.

Suppose that a narrow strip $AA'B'B$ has constant width d and rotates about the axis OO' (fig. 3.39). Denoting the volume of the resulting body by V, we may apply Guldin's first theorem to write

$$V = 2\pi R_{C'} S_{AA'B'B}, \qquad \frac{V}{d} = 2\pi R_{C'} \frac{S_{AA'B'B}}{d},$$

where C' is the center of gravity of the figure $AA'B'B$. Let us now fix the arc AB and begin to decrease d. We will then have

$$\lim_{d \to 0} \frac{V}{d} = 2\pi \left(\lim_{d \to 0} R_{C'} \right) \left(\lim_{d \to 0} \frac{S_{AA'B'B}}{d} \right). \qquad (3.66)$$

But as d becomes very small, the following approximations hold:

$$V \approx S_{AB}d, \qquad S_{AA'B'B} \approx l_{AB}d,$$

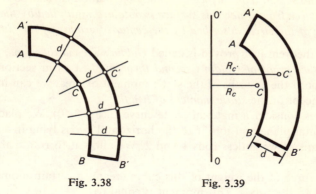

Fig. 3.38   Fig. 3.39

where S_{AB} is the area of the surface obtained by the rotation of the arc AB and l_{AB} is the length of this arc. From these relations we conclude that

$$\lim_{d \to 0} \frac{V}{d} = S_{AB}, \qquad \lim_{d \to 0} \frac{S_{AA'B'B}}{d} = l_{AB},$$

and equation (3.66) takes the form

$$S_{AB} = 2\pi \left(\lim_{d \to 0} R_{C'} \right) l_{AB}. \tag{3.67}$$

Furthermore, since the strip $AA'B'B$ has constant width, for very small d its center of gravity will be close to the center of gravity of the arc AB. Therefore, as a limit we have

$$\lim_{d \to 0} R_{C'} = R_C, \tag{3.68}$$

where C is the center of gravity of the arc AB. Substituting (3.68) into (3.67), we have

$$S_{AB} = 2\pi R_C l_{AB},$$

or, dropping the index AB,

$$S = 2\pi R_C l. \tag{3.69}$$

Equation (3.69) expresses the following theorem:

THEOREM 3.3. *The surface of a body of rotation is equal to the length of the curve from which this surface is obtained multiplied by the circumference of the circle described by the center of gravity of the curve.*

The theorem just proved is called *Guldin's second theorem.*

Another proof of Guldin's second theorem. Just as in section 3.6 we introduced the concept of the static moment of area, we can introduce the concept of the *static moment of length.*

Let us consider a material plane curve L (fig. 3.40). We place it in a horizontal plane and join it to the horizontal z-axis lying in this plane by means of weightless rods 1 and 2. We allow it to rotate about this axis.

The force of the weight of this curve creates a certain moment with respect to the z-axis. This moment is equal to

$$M_z(P) = PR_C,$$

where P is the weight of the curve L and R_C is the distance from its center of gravity to the z-axis. But

$$P = \gamma l$$

where γ is the specific weight of the curve and l is its length. Consequently,

$$M_z(P) = \gamma l R_C.$$

The product $l R_C$ is called the static moment of the length l with respect to the z-axis. Denoting it by $M_z(l)$, we can write

$$M_z(l) = l R_C. \tag{3.70}$$

The static moment of length is equivalent to the moment created by the weight of a curve when $\gamma = 1$.

If we divide the curve into parts $l_1, l_2, \ldots, l_n$, we will then have

$$M_z(l) = M_z(l_1) + M_z(l_2) + \cdots + M_z(l_n). \tag{3.71}$$

Equation (3.71) allows one to prove Guldin's second theorem without difficulty.

We will use a formula for the surface area of the side of a truncated cone to compute the area of the surface generated by the segment AB (fig. 3.41). This yields the equation

$$S_{AB} = \pi(R_1 + R_2)l_{AB}.$$

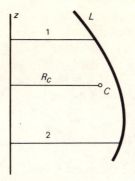

Fig. 3.40   Fig. 3.41

This may be written in the form

$$S_{AB} = 2\pi l_{AB} \frac{R_1 + R_2}{2},$$

and since

$$\frac{R_1 + R_2}{2} = R_C,$$

we have that

$$S_{AB} = 2\pi l_{AB} R_C,$$

that is,

$$S_{AB} = 2\pi M_z(l_{AB}), \tag{3.72}$$

where $M_z(l_{AB})$ is the static moment of the segment AB with respect to the axis of rotation.

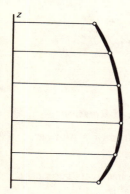

Fig. 3.42

Let us now substitute an arbitrary plane curve for the segment AB (fig. 3.42). We approximate it by a broken line consisting of n segments and suppose that n goes to infinity and that the lengths of the segments converge to zero. We then have

$$S = \lim_{n \to \infty} (S_1 + S_2 + \cdots + S_n), \tag{3.73}$$

where $S, S_1, S_2, \ldots, S_n$ are the areas of the surfaces obtained by rotating the given curve and each segment approximating the curve. The areas $S_1, S_2, \ldots, S_n$, however, can be represented in the form (3.72), which permits us to write (3.73) as

$$S = 2\pi \lim_{n \to \infty} [M_z(l_1) + M_z(l_2) + \cdots + M_z(l_n)],$$

that is,

$$S = 2\pi M_z(l), \tag{3.74}$$

where $M_z(l) = \lim_{n \to \infty} [M_z(l_1) + \cdots + M_z(l_n)]$ is the static moment of the given curve with respect to the z-axis. Substituting (3.70) and (3.74), we can now write

$$S = 2\pi R_c l.$$

This equation expresses Guldin's second theorem.

3.10. The Surface of a Sphere

Suppose that a semicircle is rotated about its diameter. Applying Guldin's second theorem, we may write

$$S = 2\pi \cdot OC \cdot \pi R$$

where O is the center of the circle and C is the center of gravity of the semicircle (fig. 3.11). Furthermore, we derived in equation (3.22) that

$$OC = \frac{2R}{\pi},$$

and, therefore,

$$S = 2\pi \frac{2R}{\pi} \pi R.$$

Consequently, the surface area of a sphere is equal to

$$S = 4\pi R^2.$$

3.11. The Surfaces of Certain Other Bodies of Rotation

Using Guldin's theorem, we can compute the area of a number of surfaces of rotation. Let us consider several examples.

1. *A torus.* Since the center of gravity of a circle lies at its geometric center, the surface area of a torus (fig. 3.27) is equal to

$$S = 2\pi R \cdot 2\pi r = 4\pi^2 Rr.$$

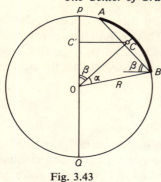

Fig. 3.43

2. *A spherical strip.* This surface is obtained by the rotation of the material arc AB about the diameter PQ (fig. 3.43). From Guldin's theorem we have

$$S = 2\pi \cdot C'C \cdot l,$$

where C is the center of gravity of the arc AB and l is its length. Furthermore, as is clear from the sketch,

$$C'C = OC \cdot \sin \beta \quad \text{and} \quad l = 2R\alpha.$$

Therefore,

$$S = 2\pi \cdot OC \cdot \sin \beta \cdot 2R\alpha.$$

Equation (3.21), however, tells us that $OC = R(\sin \alpha / \alpha)$. Consequently,

$$S = 2\pi R \frac{\sin \alpha}{\alpha} \sin \beta \cdot 2R\alpha = 2\pi R \cdot 2R \sin \alpha \sin \beta.$$

Since

$$2R \sin \alpha = l,$$

we now have

$$S = 2\pi R \cdot l \sin \beta.$$

Again referring to the sketch, we note that the second factor of this product is equal to the *altitude* of the spherical strip (that is, the projection of the chord AB onto the diameter PQ). Denoting this altitude by H, we finally obtain the formula

$$S = 2\pi R H.$$

3. Suppose that the square shown in figure 3.30 rotates about the axis OO'. The surface area of the resulting solid is equal to

$$S = 2\pi R_c l = 2\pi \left(a + \frac{a}{2} \right) \cdot 4 \frac{a\sqrt{(2)}}{2} = 6\pi \sqrt{(2)} a^2.$$

4. Guldin's theorem allows one to determine the center of gravity of certain curves. For example, knowing the surface area of a sphere, we can find the center of gravity of a semicircle. In the same way, by assuming

the formula for the surface area of a spherical strip, we can easily derive the center of gravity of a circular arc. This permits us to compute the area of the surface formed by the rotation of this arc about an arbitrary axis. In this way, in particular, it is possible to find the surface area of the body shown in figure 3.29.

3.12. Conclusion

The proofs presented in this book might suggest certain questions.

For example, one might ask whether or not we have used circular reasoning in any of the arguments. In the proof of Pythagoras' theorem we assumed the law of moments. This law is derived in physics with the help of certain physical and also *geometric* concepts. We can therefore ask whether or not Pythagoras' theorem was used in its derivation. Fortunately, an analysis of the usual derivation of the law of moments shows that it is based only on the axioms of statics and on certain theorems of similar triangles. In this instance, at least, we have avoided circular reasoning in the proof. We need to say the same thing about the other physical laws which we have assumed in this book: We have not assumed a law that is based on any of the theorems which were proved by using that law.

We might ask a second question: To what extent are the idealizations which we have used on various occasions permissible? In the third chapter, for example, we began with the concept of a line having weight but no thickness, which is admittedly impossible. To answer this question, we should point out that these idealizations are essentially the same as those used in geometry, where one speaks of a point "without length or width," and of a line "without thickness." A line having weight but not thickness is an abstraction of the same kind, arising from the representation of a thin curved rod with a definite weight, but a thickness so slight that it can be neglected. In this respect, it is possible to make further abstractions, and rather than using weight as the central property, assign to a line some other physical property, such as flexibility or elasticity. In this sense it would, for example, be possible to speak of a line having no thickness but having elastic properties. A prototype of such a line would be a thin rubber filament.[6]

---

6. The idea of a *flexible* line occurs in V. A. Uspenskii's book, *Some Applications of Mechanics to Mathematics* (New York: Blaisdell Publishing Company, 1961). The concept of an *elastic* line occurs in L. A. Lyusternik's book, *Shortest Paths: Variational Problems* (London: Pergamon Press, 1964) [translated and adapted by the Survey of Recent East European Mathematical Literature]. In these books these concepts are used for the proof of certain geometric theorems.

We might, finally, ask a third question: Do we have a right to use such nongeometric axioms as the rule of the parallelogram of forces or the postulate on the impossibility of perpetual motion? Since we introduce nongeometric objects (such as forces), however, we must introduce axioms describing the properties of these objects. Therefore, the use of nongeometric axioms in this case is natural. We can say that the proofs presented in this book are based on a system of concepts and postulates from the realm of mechanics, rather than the usual system from the realm of geometry. The fact that we are able to prove purely geometric theorems from these unusual postulates testifies to the consistency of our ideas of the physical world.